The Flock Book of Improved Leicester Sheep
Volume 1

by Improved Leicester Sheep Breeders Association

with an introduction by Jackson Chambers

Self Reliance Books

Get more historic titles on animal and stock breeding, gardening and old fashioned skills by visiting us at:

http://selfreliancebooks.blogspot.com/

Introduction

I am pleased to present yet another practical title on breeding and raising livestock.

The work is in the Public Domain and is re-printed here in accordance with Federal Laws.

As with all reprinted books of this age that are intended to perfectly reproduce the original edition, considerable pains and effort had to be undertaken to correct fading and sometimes outright damage to existing proofs of this title. At times, this task is quite monumental, requiring an almost total "rebuilding" of some pages from digital proofs of multiple copies. Despite this, imperfections still sometimes exist in the final proof and may detract from the visual appearance of the text.

I hope you enjoy reading this book as much as I enjoyed making it available to readers again.

Jackson Chambers

The Celebrated Blackfaced Ram "CAIRNTABLE,"
Bred by CHARLES HOWATSON, of Dornel, and sold for **SIXTY POUNDS,**
To J. N. FLEMING, of Keil   July, 1870

2

CONTENTS.

EDITING COMMITTEE.

CRANSWICK, J.

MARSHALL, W.

SIMPSON, J. J.

USHER, F.

THE IMPROVED LEICESTER SHEEP BREEDERS' ASSOCIATION.

PRESIDENT:

Sir CHARLES LEGARD, Bart., Ganton Hall, York.

VICE-PRESIDENT:

HASSELL POAD ROBINSON, Esq., Carnaby House, Hull.

HON. MEMBERS:

Lord MIDDLETON, Birdsall House, York.

Sir CHARLES LEGARD, Bart., Ganton Hall, York.

Sir JAMES WALKER, Bart., Sand Hutton, York.

W. H. ST. QUINTIN, Esq., Scampston Hall, Rillington, York.

W. BETHELL, Esq., Derwent House, Norton.

R. MEDFORTH, Esq., J.P., West End House, Bridlington.

Messrs. BECKETT & Co., Bankers, Driffield.

W. PEARSON, Esq., East Riding Bank, Driffield.

E. J. SMITH, Esq., York City & County Bank, Bridlington.

COMMITTEE:

APPLEBY, R. B.	MARSHALL, W.
CRANSWICK, J.	MEGGINSON, W.
CROMPTON, T.	SIMPSON, J. J.
HARLAND, T. C.	SINGLETON, J.
JACKSON, G.	TRIFFITT, W. K.
LETT, J.	USHER, F.

SECRETARY AND TREASURER:

J. CRUST, Exchange Street, Driffield.

AUDITOR:

WILFRED PEARSON (Messrs. Beckett & Co.).

BANKERS:

MESSRS. BECKETT & CO.

RULES OF THE ASSOCIATION.

1. The name of the Association is "THE IMPROVED LEICESTER SHEEP BREEDERS' ASSOCIATION."

2. The Association is established for the following objects, viz.:—

(a) The encouragement of keeping up the breeding of the Improved Leicester Sheep at home and abroad, and the maintenance of their constitution and character.

(b) The establishment and publication of a Flock Book of recognised Pure-bred Sires, which have been used, and the annual registration of the Pedigrees of such Sires as are proved to the satisfaction of the Committee to be eligible for entry.

(c) The annual compilation and publication of a statement of transactions connected with the Breed, such as particulars relating to Shows, Sales, and other necessary information.

(d) The obtaining Classes and additional Prizes at various Shows and the recommendation of the judges.

(e) The investigation of cases of doubtful and suspected Pedigrees.

(f) The undertaking of the arbitration upon and settlement of disputes, and questions relating to or connected with the Improved Leicester Sheep, and the breeding and the sale thereof, and for other subsidiary purposes.

(g) The doing of all such other lawful things as are incidental or conducive to the attainment of the above objects or any of them.

3. Any person desirous of becoming a member of this Association in future shall be proposed by one member of the Association and seconded by another member of the Association, and elected by a majority of the members present at a Committee or General Meeting of the Association.

4. The rights and privileges of every member of the Association shall be personal to himself, and shall not be transferable or transmissible by his own act or by operation of law.

5. Every member of the Association shall, provided he has paid his annual subscription for the current year, be entitled to a copy of the publication of the Association without further charge.

6. Every member on joining the Association shall pay to the fund of the Association an entrance Fee of Two Pounds, and shall also pay an annual subscription of One Pound if his Flock of Ewes exceeds in number 150, and of Ten Shillings if his Flock of Ewes shall not exceed that number.

7 The Annual Subscription shall be payable in advance on the first day of January in each year, unless the member who would otherwise have been liable to pay the same shall give notice in writing to the Secretary before that date, of his intention to withdraw from the Association.

8. No member shall withdraw from the Association without being responsible for his arrears.

9. The management of the business of the Association shall be conducted by the Committee, and the President or Chairman for the time being of committee and other meetings of this Association shall have the usual casting vote. Four members of the Committee to form a quorum.

The terms of membership shall be:—

Entrance Fee £2.

Annual Subscription £1 when the Flock of Ewes exceeds 150, and 10s. when the Flock of Ewes does not exceed that number.

Flock Book Entry Fees.

The charge to Members for inserting. Rams in the Flock Book shall be: for every Ram 1s., to non-members for every Ram 5s.

Extracts from Regulations for Entering Rams.

1. Rams to be eligible for entry must be named. and must have been used in the Flock of the person entering the same. No name in the Flock Book can be duplicated.

2. No Ram shall be eligible for entry unless bred by a Breeder whose Ewes are from a Flock which is accepted for entry in any forthcoming volume.

5. The statement of the Breeding of Rams entered must be verified by the Signature of the Breeder.

7. No Ram shall be eligible for entry unless the Name of the Sire and the Breeder of the Sire be given.

8. No Ram is eligible for entry whose Sire and Dam cannot be found to be of the Improved Leicester Breed at the time of entry.

9. Applications for registration of Sheep are received only on the understanding that the applicant possesses a Flock of Improved Leicester Ewes, and agrees to the inspection of his Flock if necessary, and decision of the inspection committee on the same, and further agrees to bear half of the expenses of inspection.

PREFACE.

THE necessity of establishing a Flock Book in the interests of Breeders of Improved Leicester Sheep had long been apparent; for whilst various breeds of Sheep identified with other parts of the country have been preserved from extinction by the enterprise of Flock Masters, there seemed to be a possibility that the peculiar characteristics of the celebrated Improved Leicester would be lost if infused into other Flock Books. The desirability of regulating the breed by a recognised standard had frequently suggested itself to those most interested in this type of Sheep. This idea did not take root very rapidly; fortunately, however, the right men were found at last who were willing to co-operate in forwarding the movement by establishing a Flock Book on a sound and proper basis; and at a meeting held at the Keys Hotel, Driffield, on Wednesday the 15th day of February, 1893, a resolution was unanimously adopted to establish a permanent record of the Improved Leicester, and to form a Society under the title of " The Improved Leicester Sheep Breeders' Association," the objects of such Society being fully explained in the Articles of Association.

The Improved Leicester has during the last few years made rapid strides towards perfection, and come most prominently to the front. As their name implies, they are descended from the original Leicester, which is regarded as the most important of our long-woolled Breeds, arriving early at maturity and possessing great aptitude to fatten, points which have caused them to be more largely used than any other in crossing and improving other Breeds of Sheep. By continuous and judicious crossing with other sires of large size and heavy fleeces, a class of Sheep has been produced of corresponding proportions, with a fullness of wool, yet retaining the original propensity to fatten. They are very hardy and well adapted for any climate or soil; during the severe winter months being folded on turnips in the open fields on the bleak Wolds of Yorkshire, where they feed quicker than any other class of Sheep that have been wintered on the same situation, requiring less artificial food, and with a minimum proportion of loss; they are also remarkably sound in their feet, and but seldom attacked by what is generally termed "foot-rot." This hardiness of constitution is very desirable in any class of Sheep, wherever situated, and is of special importance in the case of those reared in exposed situations, where natural food may at times be scarce, and artificial substitutes not easily procurable.

Not only are the Improved Leicesters a well-constitutioned class of Sheep, but good breeders, having for a long time enjoyed a reputation as the very best on the Yorkshire Wolds. They are splendidly adapted for crossing with Colonial and Foreign Sheep, and can be specially recommended for that purpose. In regard to Wool the Improved Leicester is very wealthy, having frequently been known to produce Fleeces of clean washed Wool weighing from 21 up to 28lbs., and the coat is of a beautiful texture. They are upstanding, a good size, exceptionally full in the neck and shoulders, the chest broad and deep, back broad and firm to the touch, and quarters of a good length. The Sheep attain to a great weight, records showing that they have turned the scales at 240lbs.

Yorkshire Breeders of Leicester Sheep have been eminently successful in the Show Yards, and there are Flocks in the County which have maintained their high qualities for over a century.

The Improved Leicester will doubtless come into greater favour when its excellent properties are more widely known, and the founders of the Association anticipate an accession of foreign trade as one result of the publication of this their first Volume.

Foreign and Colonial Breeders rightly recognise the value attaching to Pedigree. English Flock Masters, in order to ensure an export trade, must satisfy the requirements of their customers in this particular; and the fact that a Flock Book has been launched in their interests, will, it is hoped and believed, act as an impetus to Breeders to endeavour to produce the very best animals of their kind, and so maintain the high repute to which the Improved Leicester has attained.

Driffield, June, 1893.

THE IMPROVED LEICESTER
SHEEP BREEDERS' ASSOCIATION.

PARTICULARS

OF THE

FLOCKS OF MEMBERS

BY WHOM RAMS HAVE BEEN ENTERED

FOR THIS VOLUME.

ROBERT BAKER APPLEBY,
WILSTHORPE, BRIDLINGTON.

Flock No. 7. Ewes put to the Ram, 200.

This Flock came into the Owner's possession in 1890 from his late father, which had been bred by him for 45 years. During that time Rams had been used from improved Leicester Breeders, including the late Mr. W. ANGAS, Neswick, Mr. RAPER, Hunmanby, lately from Mr. J. MEDFORTH, Gransmoor, Mr. F. JOHNSON, Brigham, and Mr. J. J. SIMPSON, Pillmoor.

The Sires used recently: "Carnaby 2nd.," (15), and "Carnaby Lord," (16), bred by Mr. H. P. ROBINSON, 1890-1889; "Leckonfield," (60), bred by Mr. R. FISHER, 1891; and "Lowthorpe," (66), bred by Mr. T. CROMPTON, 1891.

EDWARD THOMAS BARBER,

HAVERFIELD, PATRINGTON, HULL.

––––––

Flock No. 12.   Ewes put to the Ram, 130.

This Flock has been in the Owner's possession 2 years from Ewes and Shearlings purchased from Mr. H. H. STAVELEY, Southburn, Mr. WILLIAM MEGGINSON, Towthorpe, and Mr. ROBERT MEGGINSON, Garton Field.

Last year "Nightingale," (71), bred by Mr. J. THOMPSON WAULDBY, 1890; "Swallow," (84), and "Swift," (85), bred by Mr. SWALLOW, Horkstow Villa, 1890, were used in the flock.

––––––

ABRAHAM CLARKSON,

EASTLANDS, HUGGATE, YORK.

––––––

Flock No. 22.   Ewes put to the Ram, 220

This Flock has been in the Owner's possession 5 years, and is from the Flocks of Mr. J. BROWN, North Field, Huggate, and Mr. W. CLARKSON, Painslack.

Rams used in the Flock: "William," (98), bred by Mr. W. MARSHALL, Cottam, 1889; "Ernest," (36), bred by Mr. E. F. JORDAN, Eastburn, 1890; and "James," (59), bred by Mr. J. R. SINGLETON, Greenwick, 1891.

LEONARD CLARKSON,

HUGGATE LODGE, WETWANG, YORK.

Flock No. 16.   Ewes put to the Ram, 150.

This Flock has been in the possession of the Owner 11 years, was formed by the Owner's uncle, the late Mr. GEORGE CLARKSON, who was a Breeder 45 years. Since the origin of the Flock Rams have been used from the late Sir TATTON SYKES, Bart., Mr. J. STAVELEY, Southburn, and Mr. T. CROMPTON, Lowthorpe, and lately from Mr. H. P. ROBINSON, Carnaby, Mr. R. FISHER, Leckonfield, Mr. H. H. STAVELEY, Southburn, and Mr. W. MARSHALL, Cottam.

Rams used last year: "Crop Ear," (26), "Black Foot," (11), and "Eastburn," (32), bred by Mr. E. F. JORDAN, Eastburn.

WHARRAM CLARKSON,

PAINSLACK, WETWANG, YORK.

Flock No. 20.   Ewes put to the Ram, 300.

This Flock descended to the Owner from his late father 9 years ago, and had previously been in existence 35 years. During that time Rams had been used from the Flocks of Mr. J. R. SINGLETON, Givendale, Mr. J. STAVELEY, Southburn, and Mr. H. P. ROBINSON, Carnaby House.

Rams used last year: "Wanton," (95), "Wrangler," (102), bred by Mr. H. P. ROBINSON, Carnaby House, 1891, 1889; "Watnot," (96), bred by J. R. SINGLETON, Greenwick, 1889; "Westburn," (97), bred by Mr. E. F. JORDAN, Eastburn, 1890, and "Walter," (94), bred by the Owner, 1890.

B

JOHN CRANSWICK,

FIELD HOUSE, HUNMANBY.

Flock No. 3. Ewes put to the Ram, 300.

This Flock has been in the Owner's possession 13 years, being established from Ewes bought of the late Mr. REASTON, Carnaby, Mr. KEITH, Carnaby, and Mr. ETHERINGTON, Spring Dale, also from Mr. T. LEE, Field House, Hunmanby. Those from Mr. LEE had been bred by the late Mr. J. SIMPSON, Hunmanby, that Flock being in existence nearly 70 years.

Rams used in the Flock: "Sir Charles," (79), bred by Mr. H. Dudding, Riby Grove, 1891; "Woldsman, (100), bred by Mr. H. SMITH CROPWELL BUTLER, 1891; and "Woldranger," (101), bred by Mr. R. FISHER, Leckonfield, 1890.

THOMAS CROMPTON,

HOUNDALES, NAFFERTON.

Flock No. 13.

Bought 90 Ewes and 40 Shearlings from his father's Exors. in 1893. This Flock had been in the possession of the late Mr. T. CROMPTON, Lowthorpe, for 30 years. During that time Rams of the finest quality and the greatest size have been used.

Rams used in the Flock last year: "Barff Echo," (10), bred by Mr. T. C. DIXON, Brandesburton Barff, 1891; "Carnaby Swell," (23), bred by Mr. H. P. ROBINSON, 1889; "Leckonfield Chief," (61), and "Leckonfield Junior," (62), bred by Mr. R. FISHER, Leckonfield, 1890 and 1891; "Nocton Active," (72), bred by Messrs. WRIGHT, Nocton Heath, 1890; "Lowthorpe Black Foot," (67), "Wold Champion," (99), bred by Mr. T. CROMPTON'S Exors., 1891 and 1890; and "Watchful," (93), bred by Mr. C. CLARK, Scopwick, 1891.

FRANCIS A. HAMILTON, ESQ.,

Warren Farm, Ganton, York.

Flock No. 4.   Ewes put to the Ram, 420.

The Owner has been in possession of this Flock since 1883, and which had been judiciously bred for 50 years previously, from the Flocks of the late Mr. Key, Musley Bank, Mr. Danby, High Mowthorpe, Mr. Hicks, of Little Givendale. Rams have also been used from the Flocks of Mr. T. C. Dixon, Brandesburton Barff, Mr. H. P. Robinson, Carnaby, Mr. J. Lett, Scampston, Mr. F. Danby, High Mowthorpe, Mr. Spink, Hunmanby, and Mr. R. Fisher, Leckonfield.

Last year, Rams used were: "Carnaby First," (14), bred by Mr. H. P. Robinson, Carnaby, 1890; "Charles the First," (25), also "Freeman," (38), and "Simon," (77), bred by Mr. T. C. Dixon, Brandesburton Barff, 1891; "One Eye," (73), bred by the Owner, 1890; and "Yorkshireman," (103), bred by Mr. F. Danby, 1890.

THOMAS CHARLES HARLAND,

Holme Wold House, Hull.

Flock No. 10.   Ewes put to the Ram, 170.

This Flock came into the Owner's possession in 1883, from his late father, it has been in existence 50 years. Bred from the Flocks of Mr. J. R. Singleton, Givendale, and Mr. R. Fisher, Leckonfield.

Rams used: "Lord Leckonfield," (64), "Lord Leckonfield Second," (65), bred by Mr. R. Fisher, 1889, 1891; and "Middlethorpe," (70), bred by Mr. Usher, Middlethorpe, 1890.

GEORGE JACKSON,

EAST GATE HOUSE, SEAMER, YORK.

Flock No. 6.   Ewes put to the Ram, 100.

Flock commenced in 1888 from the Flocks of Mr. HENRY JACKSON and Exors. of the late Mr. THOMAS JACKSON, Deepdale, which had been bred 50 years previously from good bred Flocks.

Last year " Dash," (28), and " Daylight," (29), bred by Mr. J. LETT, Scampston, 1891, were used in the Flock.

JOHN LETT,

CLEVELAND STUD FARM, SCAMPSTON, RILLINGTON, YORK.

Flock No. 9.   Ewes put to the Ram, 220.

This Flock has been in the Owner's possession 13 years, taken from his late father. The Flock had been in existence since 1863, was descended from the Flocks of the late Sir TATTON SYKES, Bart., and the late Mr. R. TOPHAM, Scampston, from which several prize winners have been bred. Lately Rams have been used from Mr. J. CASSWELL, Pointon, Mr. T. CROMPTON, Lowthorpe, and Mr. T. C. DIXON, Brandesburton Barff.

Rams used in the Flock: " A One," (7), " Dandy," (17), and " Scampston," (76), bred by the Owner, 1890, 1888, and 1886 ; also Mr. T. C. DIXON'S " No. 9," (30), 1891.

WILLIAM MARSHALL,

COTTAM HOUSE, DRIFFIELD.

Flock No. 5.   Ewes put to the Ram, 400.

This Flock was taken in 1889 from his father's, Mr. JAMES MARSHALL'S, Blanche, who took his Flock from Mr. THOMAS MARSHALL, of Blanche, in 1865, which had then been in the Marshall family since 1822. Since the commencement of the Flock Rams have been used from the late Sir TATTON SYKES, Bart., Mr. WILEY, Mr. STAMPER, Mr. LEIGHTON, and Mr. J. STAVELEY, lately from Mr. FISHER and Mr. H. P. ROBINSON.

Last year's Rams were "Eastburn Surprise," (33), "Eastburn Swell," (34), "Eastburn Victor," (35), bred by Mr. E. F. JORDAN, Eastburn, 1890, 1890, 1891; "Catterick King," (24), and "Topsman," (88), bred by Mr. T. H. HUTCHINSON, Catterick, 1890, 1891; "Flockmaster," (37), and "Toby," (86), bred by the Owner, 1890, 1891.

WILLIAM DEWSBURY MEGGINSON,

TOWTHORPE, WHARRAM, YORK.

Flock No. 19.   Ewes put to the Ram, 350.

This Flock has been in the Owner's possession 20 years, and came to him from his late father, Mr. D. MEGGINSON. This Flock has been in the family over 100 years, where high class Rams have always been used, they being from the Flocks of the late Sir TATTON SYKES, Mr. J. SMITH, Marton Lodge, Mr. T. CROMPTON, Lowthorpe, Mr. R. KIRBY, Watton Carr, and Mr. J. R. SINGLETON, Greenwick.

Rams used last year "Towthorpe," (89), "Towthorpe 2nd," (90), Towthorpe 3rd," (91), Towthorpe 4th, (92), bred by Mr. E. F. JORDAN, Eastburn, 1889, 1889, 1889, and 1891; "Holderness," (49), and "Improver," (58), bred by Mr. R. FISHER, Leckonfield, 1891, 1891.

GEORGE MOSEY

SKERNE, DRIFFIELD.

Flock No. 8.   Ewes put to the Ram, 140.

Flock commenced 15 years ago, made by purchases from the late Mr. W. HOLTBY, Ruston Parva, Mr. G. WALMSLEY, Rudston, Mr. J. CARRICK, Millingdale, Messrs. ALLANSON, Kendall, Mr. B. JOHNSON, Frodingham Bridge, Mr. J. MILNER, Kilham, in 1893 from Mr. F. DAY, Carnaby, all these Breeders having well established Flocks.

Rams used in the Flock: "Skerne," (80), and "Skerne No. 1," (81), bred by Mr. T. C. DIXON, Brandesburton Barff, 1891; "Skerne No. 2," (82), bred by Mr. H. P. ROBINSON, Carnaby, 1891; and "Skerne No. 3," (83), bred by Owner, 1891.

JOHN NIGHTINGALE,

STEPNEY HOUSE, BRIDLINGTON.

Flock No. 18.   Ewes put to the Ram, 200.

Mr. NIGHTINGALE has been a breeder 7 years. Ewes have been bought from the late Mr. SMITH, Marton Lodge, Mr. B. JOHNSON, Frodingham, Mr. ROBINSON, Harpham, the late Mr. T. CROMPTON, Lowthorpe, and Mr. JACKSON, Seamer, those being Breeders of high class Sheep.

Rams used last year: "Huntow," (54), "Huntow 2nd," (55), "Huntow 3rd," (56), "Huntow 4th," (57), bred by Mr. H. P. ROBINSON, Carnaby, 1891.

HASSELL POAD ROBINSON,
CARNABY HOUSE, HULL.

Flock No. 1.   Ewes put to the Ram, 270.

This Flock has been in existence over 150 years, taken by the present Owner from his uncle the late Mr. HENRY ROBINSON, Carnaby House, who inherited it from his father, Mr. GEORGE ROBINSON, Carnaby House, who was a most successful exhibitor of Sheep at the Royal, Yorkshire, and other principal Shows, having won over 400 prizes; he sold several at high prices which went to our Colonies, also several into Nottingham, Lincolnshire, and Yorkshire; it was established by his father, Mr. THOMAS ROBINSON. Rams have been introduced into this Flock from the following renowned Breeders, viz.: Messrs. WHITE, STUBBINS, EARNSHAW, FARROW, SHARPE, ROSE, CHAMPION, BUCKLEY, BAKEWELL, STONE, BURGESS, SANDY, CRESSWELL, GREEN, and the late Sir TATTON SYKES, Bart., KIRKHAM, DUDDING, CLARKE, CASWELL, SMITH, and WRIGHT.

Last year Rams used : "Carnaby No. 1," (17), bred by Mr. T. CASSWELL, Pointon, 1888; "Carnaby No. 2," (18), "Carnaby No. 3," (19), and "Carnaby No. 5," (21), bred by the Owner, 1890, 1890, and 1891; "Carnaby No. 4," (20), bred by Mr. R. WRIGHT, Nocton Heath, 1891, and "Carnaby No. 6," (22), bred by Mr. NEEDHAM, Hultoft Grange, 1891.

JOHN JORDAN SIMPSON,
PILLMOOR HOUSE, HUNMANBY R.S.O.

Flock No. 2.   Ewes put to the Ram, 165.

This Flock has been in the Owner's possession 39 years, taken from his late father, Mr. JOHN SIMPSON, Hunmanby, and the late Mr. G. SIMPSON, Marton. It had then been in existence over 30 years. Rams of the Improved Leicester type have been used in the Flock from the late Mr. G. WALMSLEY, Rudston; also from Mr. T. C. DIXON, Brandesburton Barff, Mr. H. P. ROBINSON, Carnaby House, Mr. R. FISHER, Leckonfield, and Mr. T. SPINK, Hunmanby.

Rams used last year: "Ameer," (6), bred by Mr. T. SPINK, Hunmanby, 1890; "Sindia," (78), and "Zemindar," (104), bred by the Owner, 1891.

JOHN SINGLETON,

GREAT GIVENDALE, POCKLINGTON.

Flock No. 15.   Ewes put to the Ram, 180.

The Owner commenced his Flock of Improved Leicesters in 1889 by the purchase of 50 Ewes and 40 Shearlings from Mr. F. DANBY, High Mowthorpe; 28 Ewes from Mr. GRUNDON, Neswick; 40 Shearlings from Mr. P. HICKS, Little Givendale, and, in 1890, 40 Shearlings from Mr. P. HICKS. At the Greenwich Sale in 1892 (Mr. J. R. SINGLETON), 5 Ewes and 18 Shearlings were purchased, these are descended from Ewes bred by the late Mr. H. EDWARDS, Market Weighton, nearly 50 years ago.   Rams were used to them from the Flocks of the late Sir TATTON SYKES, Mr. WILEY, and Messrs. SANDY, STONE, and BUCKLEY.

Rams lately used in the Flock: "Greenwick," (39), "Greenwick 2nd," (40), "Greenwick 3rd," (41), "Greenwick 4th," (42), "Greenwick 5th," (43), "Greenwick 6th," (44), "Greenwick 7th," (45), "Greenwick 8th," (46), "Greenwick 9th," (47), "Greenwick 10th," (48), bred by Mr. SINGLETON, Greenwick, 1888, 1890, 1891.

THOMAS R. STORK,

THE MANOR, BRIGHAM, HULL.

Flock No. 20.   Ewes put to the Ram, 150.

This Flock has been in the possession of the Owner 7 years, obtained from the Flocks of Mr. M. LEAPER, Sledmere, Mr. FOSTER, Littlethorpe, and the late Mr. B. JOHNSON, Frodingham Bridge.   These Flocks have been in existence a number of years, being bred from Improved Leicesters.

Rams lately used: "Topper," (87), bred by the late Mr. T. CROMPTON, Lowthorpe, 1890; "Pride," (74), bred by Mr. J. DIXON, Nafferton Wold, 1891; and "Prince," (75), bred by the late Mr. B. JOHNSON, Frodingham, 1891.

JOSEPH THOMPSON,

WAULDBY, BROUGH.

Flock No. 23.   Ewes put to the Ram, 300.

This Flock came into the Owner's possession 26 years ago, descended to him from his late father, who had been a Breeder of Leicester Sheep 35 years. Rams used in the Flock from the following Breeders: the late Mr. TORR, Aylesby, Mr. R. FISHER, Leckonfield, and Mr. J. STAVELEY, Southburn, also from Mr. J. R. SINGLETON, Givendale, Mr. HARRISON, Lealholme, and Mr. KIRBY, Watton Carr.

Rams used last year: "Abbot," (1), "Admiral," (2), and "Ajax," (3), bred by R. KIRBY, Watton Carr, 1889, 1890; "Albert," (4), bred by J. GILLIATT, Dowthorpe Hall, 1889; "Alderman," (5), "Artist," (8), and "Auditor," (9), bred by the Owner, 1891.

MARTIN THOMPSON,

CARNABY, HULL.

Flock No. 17.   Ewes put to the Ram, 150.

This Flock has been in the possession of Mr. THOMPSON upwards of 20 years, during which time Rams have been used from the Carnaby Flock. In 1890 an addition was made to the Flock by the purchase of Ewes from the late Mr. ROBINSON, Harpham.

Rams used in the Flock: "Bonny Scotland," (12), "Brackendale," (13), and "Doctor," (31), bred by Mr. H. P. ROBINSON, Carnaby, 1891.

FRANK USHER,

MIDDLETHORPE, MARKET WEIGHTON.

Flock No. 11.　Ewes put to the Ram, 150.

The Owner of this Flock inherited it from his late father, Mr. W. USHER, Middlethorpe, 7 years ago, who took it from the Owner's grandfather: it has been in existence over 90 years. In 1893, 30 Ewes were purchased from Mr. J. MATHISON, Sober Hill; they are descended from the Flocks of Mr. E. F. JORDAN, Eastburn, and the late Mr. W. BROWN, Holme-on-Spalding-Moor. Rams have been used in the Flock from Mr. T. H. HUTCHINSON, Catterick, Mr. J. R. SINGLETON, Givendale, Mr. W. BROWN, Highgate, Mr. J. MATHISON, Sober Hill, and C. LEAKE, North Cliffe.

Last year: "Hunmanby," (50), "Hunmanby 2nd," (51), "Hunmanby Beau," (52), and "Hunmanby Squire," (53), bred by Mr. J. J. SIMPSON, Hunmanby, 1890, were used.

WILLIAM HODGSON WILSON,

BUTTERWICK, GANTON, YORK.

Flock No. 14.　Ewes put to the Ram, 200.

This Flock has been in the Owner's possession 27 years, taken from his father, the late Mr. WILLIAM WILSON, Butterwick, who had been a breeder over 20 years; during that time Improved Leicester Rams have been used.

Rams used in the Flock: "Legard," (63), "Limber," (68), "Lustre," (69), bred by Mr. R. RICHARDSON, Arnold Grange, 1890, 1890, 1891.

REGISTER OF RAMS.

☞ The Pedigrees of the Sires in this Register was restricted by the Committee to the naming of the Sire and the Grand Sire.

REGISTER OF RAMS.

Abbreviations used: *s.* Sire; *g.s.* Grand Sire.

ABBOT.—(1)   *Lambed* 1889.

Owner : JOSEPH THOMPSON, Wauldby, Brough.
Breeder : R. KIRBY, Watton Carr, Cranswick, Hull.

ADMIRAL.—(2)   *Lambed* 1889.

Owner : JOSEPH THOMPSON, Wauldby, Brough.
Breeder : R. KIRBY, Watton Carr, Cranswick, Hull.

AJAX.—(3)   *Lambed* 1890.

Owner : JOSEPH THOMPSON, Wauldby, Brough.
Breeder : R. KIRBY, Watton Carr, Cranswick, Hull.

ALBERT.—(4)   *Lambed* 1889.

Owner : JOSEPH THOMPSON, Wauldby, Brough.
Breeder : J. GILLIAT, Dowthorpe Hall, HULL.

———

ALDERMAN.—(5)   *Lambed* 1891.

Owner and Breeder : JOSEPH THOMPSON, Wauldby,
Brough.
s. Abbot (1).

———

AMEER.—(6)   *Lambed* 1890.

Owner : JOHN J. SIMPSON, Pillmoor House, Hunmanby.
Breeder : T. SPINK, of Hunmanby.
s. bred by R. WRIGHT, Nocton Heath.

———

A ONE.—(7)   *Lambed* 1890.

Owner and Breeder : JOHN LETT, Cleveland Stud Farm,
Scampston, Rillington, York.
s. Dandy (27).

ARTIST.—(8)   *Lambed* 1891.

> *Owner and Breeder :* Joseph Thompson, Wauldby, Brough.
>
> *s.* Abbot (1).

———

AUDITOR.—(9)   *Lambed* 1891.

> *Owner and Breeder :* Joseph Thompson, Wauldby, Brough.
>
> *s.* bred by G. M. Gale, Atwick Hall, Hull.

———

BARFF ECHO.—(10)   *Lambed* 1891.

> *Owner and Breeder :* T. C. Dixon, of Brandesburton Barff.
>
> *s.* Nocton Ram ; *g.s.* Kirkham Ram.

———

BLACK FOOT.—(11)   *Lambed* 1888.

> *Owner :* L. Clarkson, Huggate Lodge, Wetwang, York.
>
> *Breeder :* E. F. Jordan, of Eastburn.

BONNY SCOTLAND.—(12)   *Lambed* 1891.

Owner : MARTIN THOMPSON, Carnaby, Hull.
Breeder : H. P. ROBINSON, of Carnaby House.
s. bred by Mr. CLARK, of Scopwick.

———

BRACKENDALE.—(13)   *Lambed* 1891.

Owner : MARTIN THOMPSON, Carnaby, Hull.
Breeder : H. P. ROBINSON, of Carnaby House.
s. bred by Mr. CASSWELL, Pointon.

———

CARNABY 1st.—(14)   *Lambed* 1890.

Owner : F. A. HAMILTON, Esq., Warren Farm,
Ganton, York.
Breeder : H. P. ROBINSON, of Carnaby House.

———

CARNABY 2nd.—(15)   *Lambed* 1890.

Owner : R. B. APPLEBY, Wilsthorpe, Bridlington.
Breeder : H. P. ROBINSON, of Carnaby House.

CARNABY LORD.—(16)   *Lambed* 1889.

    Owner : R. B. APPLEBY, Wilsthorpe, Bridlington.
    Breeder : H. P. ROBINSON, of Carnaby House.

———

CARNABY No. 1.—(17)   *Lambed* 1888.

    Owner : H. P. ROBINSON, Carnaby House, Hull.
    Breeder : T. CASSWELL, of Pointon.

———

CARNABY No. 2.—(18)   *Lambed* 1890.

    Owner and Breeder : H. P. ROBINSON, Carnaby House,
       Hull.
      s. bred by R. WRIGHT, of Nocton Heath.

———

CARNABY No. 3.—(19)   *Lambed* 1890.

    Owner and Breeder : H. P. ROBINSON, Carnaby House,
       Hull.
      s. bred by C. CLARKE, of Scopwick.

c

CARNABY No. 4. -- **(20)**   *Lambed* 1891.

> *Owner:* H. P. ROBINSON, Carnaby House, Hull.
> *Breeder:* R. WRIGHT, of Nocton Heath.

———

CARNABY No. 5.—(21)   *Lambed* 1891.

> *Owner and Breeder:* H. P. ROBINSON, Carnaby House,
> Hull.
> *s.* bred by C. CLARKE, of Scopwick.

———

CARNABY No. 6.—(22)   *Lambed* 1891.

> *Owner:* H. P. ROBINSON, Carnaby House, Hull.
> *Breeder:* J. L. NEEDHAM, of Huttoft.

———

CARNABY SWELL.— (23)   *Lambed* 1889.

> *Owner:* T. CROMPTON, Houndales, Nafferton.
> *Breeder:* H. P. ROBINSON, of Carnaby House.

CATTERICK KING.—(24)   *Lambed* 1890.

Owner : Wm. Marshall, Cottam House, Driffield.

Breeder : T. H. Hutchinson, of Catterick.

––––––

CHARLES THE FIRST.—(25)   *Lambed* 1891.

Owner : F. A. Hamilton, Esq., Warren Farm, Ganton, York.

Breeder : T. C. Dixon, Brandesburton Barff.

s. bred by the late Mr. Vessey.

––––––

CROP EAR.—(26)   *Lambed* 1887.

Owner : L. Clarkson, Huggate Lodge, Wetwang, York.

Breeder : E. F. Jordan, of Eastburn, Driffield

––––––

DANDY.—(27)   *Lambed* 1888.

Owner : T. Crompton, Houndales, Nafferton.

Breeder : J. Lett, of Scampston.

s. bred by J. Lett, Scampston.

DASH.—(28) *Lambed* 1891.

> *Owner :* GEORGE JACKSON, East Gate House, Seamer,
> York.
> *Breeder :* J. LETT, of Scampston.
> *s.* Dandy (27).

———

DAYLIGHT.—(29) *Lambed* 1891.

> *Owner :* GEORGE JACKSON, East Gate House, Seamer,
> York.
> *Breeder :* J. LETT, of Scampston.
> *s.* Dandy (27).

———

DIXON'S No. 9.—(30) *Lambed* 1891.

> *Owner and Breeder :* T. C. DIXON, of Brandesburton
> Barff, Hull.

———

DOCTOR.—(31) *Lambed* 1891.

> *Owner :* M. THOMPSON, Carnaby, Hull.
> *Breeder :* H. P. ROBINSON, of Carnaby House.

EASTBURN.—(32)   *Lambed* 1888.

Owner : LEONARD CLARKSON, Huggate Lodge, Wetwang, York.

Breeder : E. F. JORDAN, of Eastburn.

———

EASTBURN SURPRISE.—(33)   *Lambed* 1890.

Owner : WILLIAM MARSHALL, Cottam House, Driffield.

Breeder : E. F. JORDAN, of Eastburn.

———

EASTBURN SWELL.—(34)   *Lambed* 1890.

Owner : WILLIAM MARSHALL, Cottam House, Driffield.

Breeder : E. F. JORDAN, of Eastburn.

———

EASTBURN VICTOR.—(35)   *Lambed* 1891.

Owner : WILLIAM MARSHALL, Cottam House, Driffield.

Breeder : E. F. JORDAN, of Eastburn.

ERNEST.—(36) *Lambed* 1890.

> *Owner:* ABRAHAM CLARKSON, Eastlands, Huggate, York.
>
> *Breeder:* E. F. JORDAN, of Eastburn.

———

FLOCKMASTER.—(37) *Lambed* 1891.

> *Owner and Breeder:* WILLIAM MARSHALL, Cottam House Driffield.

———

FREEMAN.—(38) *Lambed* 1891.

> *Owner:* F. A. HAMILTON, Esq., Warren Farm, Ganton, York.
>
> *Breeder:* T. C. DIXON, of Brandesburton Barff, Hull.
>
> *s.* bred by J. H. CASSWELL, of Laughton.

———

GREENWICK.—(39) *Lambed* 1888.

> *Owner:* JOHN SINGLETON, Great Givendale, Pocklington.
>
> *Breeder:* J. R. SINGLETON, of Greenwick, Pocklington.
>
> *s.* bred by J. R. SINGLETON.

GREENWICK 2nd.—(40) *Lambed* 1888.

Owner : JOHN SINGLETON, Great Givendale, Pocklington.
Breeder : J. R. SINGLETON, of Greenwick.
   s. bred by J. R. SINGLETON.

———

GREENWICK 3rd.—(41) *Lambed* 1888.

Owner : JOHN SINGLETON, Great Givendale, Pocklington.
Breeder : J. R. SINGLETON, of Greenwick.
   s. bred by J. R. SINGLETON.

———

GREENWICK 4th.—(42) *Lambed* 1888.

Owner : JOHN SINGLETON, Great Givendale, Pocklington.
Breeder : J. R. SINGLETON, of Greenwick.
   s. bred by J. R. SINGLETON.

———

GREENWICK 5th.—(43) *Lambed* 1890.

Owner : JOHN SINGLETON, Great Givendale, Pocklington.
Breeder : J. R. SINGLETON, of Greenwick.
   s. bred by J. R. SINGLETON.

GREENWICK 6th.—(44) *Lambed* 1890.

> *Owner :* JOHN SINGLETON, Great Givendale, Pocklington.
> *Breeder :* J. R. SINGLETON, of Greenwick.
> *s.* bred by J. R. SINGLETON.

GREENWICK 7th.—(45) *Lambed* 1890.

> *Owner :* JOHN SINGLETON, Great Givendale, Pocklington.
> *Breeder :* J. R. SINGLETON, of Greenwick.
> *s.* bred by J. R. SINGLETON.

GREENWICK 8th.—(46) *Lambed* 1891.

> *Owner :* JOHN SINGLETON, Great Givendale, Pocklington.
> *Breeder :* J. R. SINGLETON, of Greenwick.
> *s.* bred by J. R. SINGLETON.

GREENWICK 9th.—(47) *Lambed* 1891.

> *Owner :* JOHN SINGLETON, Great Givendale, Pocklington.
> *Breeder :* J. R. SINGLETON, of Greenwick.
> *s.* bred by J. R. SINGLETON.

GREENWICK 10th.—(48) *Lambed* 1891.

> *Owner:* JOHN SINGLETON, Great Givendale, Pock-
> lington.
>
> *Breeder:* J. R. SINGLETON, of Greenwick.
>
> s. bred by J. R SINGLETON.

HOLDERNESS.—(49) *Lambed* 1891.

> *Owner:* W. D. MEGGINSON, Towthorpe, Wharram, York.
>
> *Breeder:* R. FISHER, of Leckonfield.

HUNMANBY.—(50) *Lambed* 1890.

> *Owner:* FRANK USHER, Middlethorpe, Market Weigh-
> ton.
>
> *Breeder:* J. J. SIMPSON, of Pillmoor House.

HUNMANBY 2nd.—(51) *Lambed* 1890.

> *Owner:* FRANK USHER, Middlethorpe, Market Weigh-
> ton.
>
> *Breeder:* J. J. SIMPSON, of Pillmoor House.

HUNMANBY BARON.—(52)   *Lambed* 1890.

Owner : FRANK USHER, Middlethorpe, Market Weighton.

Breeder : J. J. SIMPSON, of Pillmoor House.

———

HUNMANBY SQUIRE.—(53)   *Lambed* 1890.

Owner : FRANK USHER, Middlethorpe, Market Weighton.

Breeder : J. J. SIMPSON, of Pillmoor House.

———

HUNTOW.—(54)   *Lambed* 1891.

Owner : JOHN NIGHTINGALE, Stepney House, Bridlington.

Breeder : H. P. ROBINSON, of Carnaby House.

———

HUNTOW 2nd.—(55)   *Lambed* 1891.

Owner : JOHN NIGHTINGALE, Stepney House, Bridlington.

Breeder : H. P. ROBINSON, of Carnaby House.

HUNTOW 3rd.—(56)   *Lambed* 1891.

Owner : JOHN NIGHTINGALE, Stepney House, Bridlington.
Breeder : H. P. ROBINSON, of Carnaby House.

———

HUNTOW 4th.—(57).   *Lambed* 1891.

Owner : JOHN NIGHTINGALE, Stepney House, Bridlington.
Breeder : H. P. ROBINSON, of Carnaby House.

———

IMPROVER.—(58)   *Lambed* 1891.

Owner : W. D. MEGGINSON, Towthorpe, Wharram, York.
Breeder : R. FISHER, of Leckonfield.

———

JAMES.—(59)   *Lambed* 1891.

Owner : ABRAHAM CLARKSON, Eastlands, Huggate,
York.
Breeder : J. R. SINGLETON, of Greenwick.

LECKONFIELD.—(60) *Lambed* 1891.

> *Owner :* R. B. APPLEBY, Wilsthorpe, Bridlington.
> *Breeder :* R. FISHER, of Leckonfield.

———

LECKONFIELD CHIEF.—(61) *Lambed* 1890.

> *Owner :* the late T. CROMPTON'S Exors., Lowthorpe.
> *Breeder :* R. FISHER, of Leckonfield.
> *s.* bred by R. WRIGHT, Nocton Heath.

———

LECKONFIELD JUNIOR.—(62) *Lambed* 1890.

> *Owner :* the late T. CROMPTON'S Exors., Lowthorpe.
> *Breeder :* R. FISHER, of Leckonfield.
> *s.* bred by the late Mr. VESSEY.

———

LEGARD.—(63) *Lambed* 1891.

> *Owner :* W. H. WILSON, Butterwick, Ganton, York.
> *Breeder :* R. RICHARDSON, of Arnold Grange, Hull.
> *s.* bred by J. TURNER, Ulceby.

LORD LECKONFIELD.—(64)   *Lambed* 1889.

> *Owner :* T. C. HARLAND, Holme Wold House, Hull.
> *Breeder :* R. FISHER, of Leckonfield.

———

LORD LECKONFIELD 2nd.—(65)   *Lambed* 1891.

> *Owner :* T. C. HARLAND, Holme Wold House, Hull.
> *Breeder :* R. FISHER, of Leckonfield.

———

LOWTHORPE.—(66)   *Lambed* 1891.

> *Owner :* R. B. APPLEBY, Wilsthorpe, Bridlington.
> *Breeder :* The late T. CROMPTON'S Exors., Lowthorpe.

———

LOWTHORPE BLACK FOOT.—(67)   *Lambed* 1891.

> *Owner and Breeder :* The late T. CROMPTON'S Exors.,
> Lowthorpe.
> *s.* bred by R. FISHER, Leckonfield.

LIMBER.—(68)   *Lambed* 1891.

Owner: W. H. WILSON, Butterwick, Ganton, York.
Breeder: R. RICHARDSON, of Arnold Grange, Hull.
s. bred by J. TURNER, Ulceby.

———

LUSTRE.—(69)   *Lambed* 1891.

Owner: W. H. WILSON, Butterwick, Ganton, York.
Breeder: R. RICHARDSON, of Arnold Grange, Hull.
s. bred by J. TURNER, Ulceby.

———

MIDDLETHORPE.—(70)   *Lambed* 1890.

Owner: T. C. HARLAND, Holme Wold House, Hull.
Breeder: F. USHER, of Middlethorpe.
s. bred by T. H. HUTCHINSON, Catterick.

———

NIGHTINGALE.—(71)   *Lambed* 1890.

Owner: E. T. BARBER, Haverfield, Patrington.
Breeder: J. THOMPSON, Wauldby, Hull.
s. bred by J. THOMPSON, Wauldby.

NOCTON ACTIVE.—(72)   *Lambed* 1890.

 Owner: Exors. of the late T. CROMPTON, Lowthorpe.
 Breeder: R. WRIGHT, of Nocton Heath.
  s. Black Spot.

———

ONE EYE.—(73)   *Lambed* 1890.

 Owner: F. A. HAMILTON, Esq., Warren Farm, Ganton.
 Breeder: H. P. ROBINSON, of Carnaby House.

———

PRIDE.—(74)   *Lambed* 1891.

 Owner: T. R. STORK, Brigham, Driffield.
 Breeder: J. DIXON, Nafferton Wold, Driffield.

———

PRINCE.—(75)   *Lambed* 1891

 Owner: T. R. STORK, Brigham, Driffield.
 Breeder: the late B. JOHNSON, Frodingham Bridge.

SCAMPSTON.—(76)　　　　　　　*Lambed* 1886.

> *Owner and Breeder:* J. LETT, Scampston, Rillington.
> *s.* bred by T. C. DIXON, Brandesburton Barff.
> *g.s.* bred by the late T. CROMPTON, Lowthorpe.

——

SIMON.—(77)　　　　　　　*Lambed* 1891.

> *Owner:* F. A. HAMILTON, Esq., Warren Farm, Ganton.
> *Breeder:* T. C. DIXON, of Brandesburton Barff.
> *s.* bred by T. CASSWELL, Pointon.

——

SINDIA.—(78)　　　　　　　*Lambed* 1891.

> *Owner and Breeder:* J. J. SIMPSON, Pillmoor House.
> *s.* bred by H. P. ROBINSON, Carnaby House.

——

SIR CHARLES.—(79)　　　　　　　*Lambed* 1891.

> *Owner:* J. CRANSWICK, Field House, Hunmanby.
> *Breeder:* H. DUDDING, of Riby Grove, Lincolnshire.
> *s.* Young Scopwith.

SKERNE.—(80) *Lambed* 1891.

Owner : GEORGE MOSEY, Skerne, Driffield.

Breeder : T. C. DIXON, of Brandesburton Barff.

———

SKERNE No. 1.—(81) *Lambed* 1891.

Owner : GEORGE MOSEY, Skerne, Driffield.
Breeder : T. C. DIXON, of Brandesburton Barff.

———

SKERNE No. 2.—(82) *Lambed* 1891.

Owner : GEORGE MOSEY, Skerne, Driffield.
Breeder : H. P. ROBINSON, of Carnaby House.

———

SKERNE No. 3.—(83) *Lambed* 1891.

Owner and Breeder : GEORGE MOSEY, Skerne, Driffield.
  s. bred by T. C. DIXON, of Brandesburton Barff.
  g.s. bred by the late J. CARRICK, Lowthorpe.

D

SWALLOW.—(84)   *Lambed* 1891

    Owner : E. T. BARBER, Haverfield, Patrington.
    Breeder : J. B. SWALLOW, Horkstow Villa, Barton.

———

SWIFT. (85)   *Lambed* 1891.

    Owner : E. T. BARBER, Haverfield, Patrington.
    Breeder : J. B. SWALLOW, Horkstow Villa, Barton.

———

TOBY.—(86)   *Lambed* 1891.

    Owner and Breeder : W. MARSHALL, Cottam House,
        Driffield.

———

TOPPER.—(87)   *Lambed* 1890.

    Owner : T. R STORK, Brigham, Driffield.
    Breeder : the late J. CROMPTON, Lowthorpe.

TOPSMAN.—(88) *Lambed* 1891.

> *Owner :* W. MARSHALL, Cottam House, Driffield.
> *Breeder :* T. H. HUTCHINSON, Catterick.

———

TOWTHORPE.—(89) *Lambed* 1889.

> *Owner :* W. D. MEGGINSON, Towthorpe, Wharram, York.
> *Breeder :* E. F. JORDAN, of Eastburn, Driffield.

———

TOWTHORPE No. 2.—(90) *Lambed* 1889.

> *Owner :* W. D. MEGGINSON, Towthorpe, Wharram, York.
> *Breeder :* E. F. JORDAN, of Eastburn.

———

TOWTHORPE No. 3. —(91) *Lambed* 1889.

> *Owner :* W. D. MEGGINSON, Towthorpe, Wharram, York.
> *Breeder :* E. F. JORDAN, of Eastburn.

TOWTHORPE Nò. 4.—(92)   *Lambed* 1891.

    Owner : W. D. MEGGINSON, Towthorpe, Wharram, York.

    Breeder : E. F. JORDAN, of Eastburn.

———

WATCHFUL.—(93)   *Lambed* 1891.

    Owner : T. CROMPTON, Houndales, Nafferton.

    Breeder : C. CLARK, Scopwick.

———

WALTER.—(94)   *Lambed* 1891.

    Owner and Breeder : W. CLARKSON, Painslack, Wetwang, York.

———

WANTON.—(95)   *Lambed* 1891.

    Owner : W. CLARKSON, Painslack, Wetwang, York.

    Breeder : H. P. ROBINSON, of Carnaby House.

WATNOT.—(96) *Lambed* 1889.

> *Owner :* W. CLARKSON, Painslack, Wetwang, York.
> *Breeder :* J. R. SINGLETON, of Greenwick, Pocklington.

———

WESTBURN.—(97) *Lambed* 1890.

> *Owner :* W. CLARKSON, Painslack, Wetwang, York.
> *Breeder :* E. F. JORDAN, of Eastburn, Driffield.

———

WILLIAM.—(98) *Lambed* 1889.

> *Owner :* A. CLARKSON, Eastlands, Huggate, York.
> *Breeder :* W. MARSHALL, Cottam House, Driffield.

———

WOLD CHAMPION.—(99) *Lambed* 1890.

> *Owner :* T. CROMPTON, Houndales, Nafferton.
> *Breeder :* Exors. of the late T. CROMPTON, Lowthorpe.
> *s.* bred by Mr. T. C. DIXON.

WOLDSMAN.—(100)   *Lambed* 1891.

Owner : J. CRANSWICK, Field House, Hunmanby.
Breeder : H. SMITH, of Cropwell Butler, Nottingham.

———

WOLDRANGER.—(101)   *Lambed* 1890.

Owner : J. CRANSWICK, Field House, Hunmanby.
Breeder : R. FISHER, of Leckonfield.

———

WRANGLER.—(102)   *Lambed* 1889.

Owner : W. CLARKSON, Painslack, Wetwang, York.
Breeder : H. P. ROBINSON, of Carnaby House.

———

YORKSHIREMAN.—(103)   *Lambed* 1890.

Owner : F. A. HAMILTON, Esq., Warren Farm, Ganton, York.
Breeder : F. DANBY, of High Mowthorpe, Wharram, York.
s. bred by T. C. DIXON, of Brandesburton Barff.

ZIMINDAR.—(104)   *Lambed* 1891.

Owner and Breeder: J. J. SIMPSON, Pillmoor House, Hunmanby.

s. bred by T. C. DIXON, of Brandesburton Barff.

NOTICE.

Certificates of Registration—One Shilling each—may be obtained of the Secretary,

JOSEPH CRUST,

Exchange Street,

Driffield.

REGULATIONS AND INSTRUCTIONS
FOR
ENTERING PEDIGREES.

1.—*Rams to be eligible for entry must be named, and must be entered by the Breeder or by the person using the same. No name which has already appeared in the Flock Book can be duplicated.*

2.—*No Ram will be eligible for entry unless bred by a Breeder whose Ewes are from a Flock which is already entered in the Flock Book or accepted for entry in any forthcoming Volume.*

3.—No Ram shall be eligible in the Flock Book unless application for entry shall be made whilst in the United Kingdom, or unless satisfactory reason be given for such application not having been so made.

4.—The Committee reserve the right of declining any entry, if so recommended by the Editing and Inspection Committees.

5.—The statement of the breeding of all Rams entered must be verified by the signature of the Breeder in column 9, allotted for that purpose.

6.—No Breeders' entry will be received unless he complies with the following conditions:—

  (*a*) **THAT THE WHOLE OF THE SIRES** used in his Flock from the date of his first registration be inserted in the Flock Book. Thus, if the entries of a Flock date back to the year 1892, **ALL RAMS USED IN THAT FLOCK FROM THAT DATE MUST BE ENTERED.**

  (*b*) If required, satisfactory proof must be given of the means by which gentlemen registering Rams have hired or purchased them.

7.—*No Ram shall be eligible for entry unless the name of the Sire and the breeder of the Sire be given.*

8.—No Ram is eligible for entry whose Sire and Dam cannot be proved to the satisfaction of the above Committees to be of the Improved Leicester Breed at the time of entry.

9.—Applications for registration of Sheep are received only on the understanding that the applicant possesses a Flock of Improved Leicester Ewes, and agrees to the inspection of his Flock, if necessary, and decision of the Inspection Committee on the same, and **further agrees to bear half the cost of the expenses of inspection.**

10.—No Breeder whose Flock is registered in either the first or subsequent volumes of the Flock Book will be allowed to enter Rams prior to the date of his last registration, except by the special permission of the Editing Committee.

11.—*Any Breeder who has introduced Ewes from any other Flock into his own since his last registration, must furnish the Secretary with particulars of numbers and breeding, within seven days of such introduction, otherwise his entries will not be considered.*

12.—Every Ram entered in the Flock Book **must have a distinct name**. If this Regulation is not observed the Editing Committee reserve to themselves the right of re-naming the Ram, after giving notice of such alteration.

13.—It will greatly aid in the preparation of the Flock Book if Breeders and others will write the pedigrees **legibly** and **correctly**, and will be careful to give **every particular required** in the form of entry. Before sending in pedigrees, their accuracy should be carefully checked. Much inconvenience, delay, and correspondence will thus be avoided.

14.—**ALL ENTRIES MUST BE MADE ON THE ASSOCIATION'S PRINTED FORM,** and must be certified by the Breeder or Owner, or his accredited agent or representative, and in cases of the **first entry the accompanying questions on the fourth page must be fully and truthfully answered.** Entries must be accompanied by the necessary fees, as follows:—

To a Member of the Society ... ... 1s. each Ram.

(If Subscription is in arrear, entries are refused.)

To a Non-Member ... ... 5s. each Ram.

PARTICULAR ATTENTION IS CALLED TO THE REGULATIONS PRINTED IN ITALICS.

15.—The printed form marked "A" to be filled up by those whose Flocks are registered in the first volume of the Flock Book.

—·⊰ INDEX. ⊱·—

OWNERS AND BREEDERS OF RAMS,

SIRES, AND GRAND SIRES.

CROMPTON, T. (The Late), Lowthorpe—66, 76, 87.

DANBY, F., High Mowthorpe, Wharram, York—103.

DIXON, J., Nafferton Wold, Driffield—74.

DIXON, T. C., Brandesburton Barff—10, 25, 30, 38, 76, 77, 80, 81, 83, 103, 104.

DUDDING, H., Riby Grove, Lincolnshire—79.

FISHER, R., Leckonfield—49, 58, 60, 61, 62, 64, 65, 67, 101.

GALE, G. M., Atwick Hall, Hull—9.

GILLIAT, J., Dowthorpe Hall, Hull—4.

HAMILTON, F. A., Warren Farm, Ganton, York—14, 25, 38, 73, 77, 103.

HARLAND, T. C., Holme Wold House, Hull—64, 65, 70.

HUTCHINSON, T. H., Catterick—24, 70, 88.

JACKSON, GEORGE., East Gate House, Seamer, York—28, 29.

JOHNSON, B. (The Late), Frodingham Bridge—75.

JORDAN, E. F., Eastburn, Driffield—11, 26, 32, 33, 34, 35, 36, 89, 90, 91, 92, 97.

KIRBY, R., Watton Carr, Cranswick, Hull—1, 2, 3.

LETT, JOHN, Cleveland Stud Farm, Scampston, Rillington, York—7, 27, 28, 29, 76.

MARSHALL, WM., Cottam House, Driffield—24, 33, 34, 35, 37, 86, 88, 98.

MEGGINSON, W. D., Towthorpe, Wharram, York—49, 58, 89, 90 91, 92.

MOSEY, GEORGE, Skerne, Driffield—80, 81, 82, 83.

NEEDHAM, J. L., Huttoft—22.

NIGHTINGALE, JOHN, Stepney House, Bridlington—54, 55, 56, 57.

RICHARDSON, R., Arnold Grange, Hull—63, 68, 69.

ROBINSON, H. P., Carnaby House—12, 13, 14, 15, 16, 17, 18, 19, 20, 21,
    22, 23, 31, 54, 55, 56, 57, 73, 78, 82, 95.

SIMPSON, JOHN J., Pillmoor House, Hunmanby—6, 50, 51, 52, 53, 78, 104.

SINGLETON, JOHN, Great Givendale, Pocklington—39, 40, 41, 42, 43, 44,
    45, 46, 47, 48.

SINGLETON, J. R., Greenwick, Pocklington—39, 40, 41, 42, 43, 44, 45,
    46, 47, 48, 59, 96.

SMITH, H., Cropwell Butler, Nottingham—100.

SPINK, T., Hunmanby—6.

STORK, T. R., Brigham, Driffield—74, 75, 87.

SWALLOW, J. B., Horkstow Villa, Barton—84, 85.

THOMPSON, JOSEPH, Wauldby, Brough—1, 2, 3, 4, 5, 8, 9, 71.

THOMPSON, MARTIN, Carnaby, Hull—12, 13, 31.

TURNER, J., Ulceby—63, 68, 69.

LIST OF MEMBERS.

Appleby, Robert Baker, Wilsthorpe, Bridlington.

Barber, Edward Thomas, Haverfield, Patrington, Hull.

Clarkson, Abraham, Eastlands, Huggate, York.

Clarkson, Leonard, Huggate Lodge, Wetwang, York.

Clarkson, Wharram, Painslack, Wetwang, York.

Cranswick, John, Field House, Hunmanby.

Crompton, Thomas, Houndales, Nafferton.

Hamilton, Francis A., Esq., Warren Farm, Ganton, York.

Harland, Thomas Charles, Holme Wold House, Hull.

Jackson, George, East Gate House, Seamer, York.

Lett, John, Cleveland Stud Farm, Scampston, Rillington, York.

Marshall, William, Cottam House, Driffield.

Megginson, William Dewsbury, Towthorpe, Wharram, York.

Mosey, George, Skerne, Driffield.

Nightingale, John, Stepney House, Bridlington.

Robinson, Hassell Poad, Carnaby House, Hull.

Simpson, John Jordan, Pillmoor House, Hunmanby R.S.O.

Singleton, John, Great Givendale, Pocklington.

Stork, Thomas R., The Manor, Brigham, Hull.

Thompson Joseph, Wauldby, Brough.

Thompson, Martin, Carnaby, Hull.

Usher, Frank, Middlethorpe, Market Weighton.

Wilson, William Hodgson, Butterwick, Ganton, York.

B. FAWCETT & CO., PRINTERS, DRIFFIELD.

J. RANDS & JECKELL'S
SHEEP AND LAMB
SHELTERING CLOTHS

Made up in 20 yard lengths, 3 feet deep, with brass eyelets and Cords, ready to fix to hurdles.

6d., 9d. and 1s. per yard.   Carriage Paid.

USED ON THE FARMS OF H.M. THE QUEEN.

[COPY.]   HENGRAVE, BURY ST. EDMUNDS, DEC. 19th, 1890.

Gentlemen,—We have much pleasure in bearing testimony to the value of your Sheep Cloths.   We think them by far the cheapest and easiest way of providing shelter for lambs against the cold winds of the spring.

We remain, yours truly,
ROBERSON & GOUGH.

Messrs. J. Rands & Jeckell, Ipswich.

[COPY.]   MANOR FARM, ROPLEY, NEAR ALRESFORD,
January 30th, 1893.

Dear Sirs,—Kindly send me to Ropley Station, 4 20-yard lengths of your Lamb Sheltering Cloths, at 9d. per yard.   That which I had of you I find wonderfully useful, both in lambing fold and also in the field. I recommend all my friends to you.

Yours faithfully,
ROBT. M. KEEN.

J. Rands & Jeckell, Ipswich.

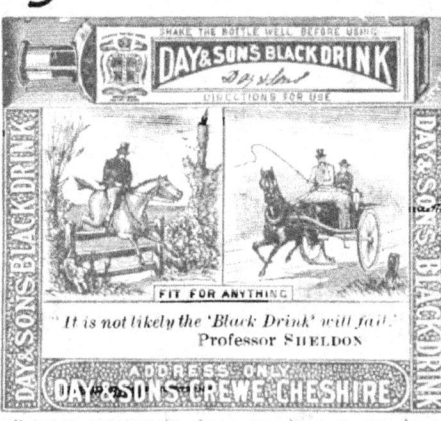